Jorge Gómez
Velssy Hernández

TEACHING MACHINE LEARNING PROGRAMMING IN PANDAS AND JUPYTER-LAB

Jorge Gómez
Velssy Hernández

TEACHING MACHINE LEARNING PROGRAMMING IN PANDAS AND JUPYTER-LAB

Beginner Level

ScienciaScripts

Imprint

Cover image: www.ingimage.com

This book is a translation from the original published under ISBN 978-3-659-70272-3.

Publisher:
Sciencia Scripts
is a trademark of
Dodo Books Indian Ocean Ltd. and OmniScriptum S.R.L publishing group

120 High Road, East Finchley, London, N2 9ED, United Kingdom
Str. Armeneasca 28/1, office 1, Chisinau MD-2012, Republic of Moldova, Europe
Printed at: see last page
ISBN: 978-620-5-65723-2

Content

SUMMARY

This book is designed for those who wish to enter the world of machine learning, i.e. beginner mode. Initially, data manipulation using dataframes in the Python programming language, embedded in the Jupyter-lab framework and Pandas, will be addressed. Data will then be extracted from csv files to be managed in dataframes. Indexing, selection and allocation, indexing in Pandas, tag-based selection, conditional selection, data allocation, summary functions, maps and grouping and sorting. Finally, we will proceed to program basic supervised learning models such as linear regression with a single variable, linear regression with multiple variables, saving and loading the training model, data management with dummy variables, separation of the training and test datasets.

ABOUT THE AUTHORS

Jorge Gómez Gómez

Systems Engineer, received a Master's degree in Telematics Engineering at the University of Cauca Colombia in 2010, PhD in Information Technology and Communications at the University of Granada Spain in 2018, Full-time professor of the Systems Engineering program - University of Cordoba, Member IEEE Branch. Research interests: Ubiquitous Computing and Advanced Services in Telecommunications. Researcher in the SOCRATES research group - University of Cordoba.

Velssy Hernández Riaño:

She is a Systems Engineer and holds a Master's degree in Telematics Engineering from the Universidad Francisco José de Caldas, Colombia. She is a Professor and Researcher in the SOCRATES research group of the Department of Systems Engineering at the University of Córdoba.

INTRODUCTION

Machine learning is a set of algorithms specialised in prediction techniques, which are used for a number of applications, ranging from medicine, astronomy, geology, biology and other sciences. Classical machine learning is classified according to the way in which an algorithm learns to be more efficient in its predictions.

Machine learning is the study of computer algorithms that can learn complex relationships or patterns from empirical data and make accurate decisions (Carleo et al., 2019; Alpaydin; El Naqa & Murphy, 2015). It is an interdisciplinary field that has close relationships with artificial intelligence, pattern recognition, data mining, statistics, probability theory, optimisation, statistical physics and theoretical computer science. Applications of machine learning include natural language processing, medical diagnostics, bioinformatics, video surveillance and financial data analysis.

Machine learning algorithms can be organised into different categories according to different principles. For example, depending on the use of the labels of the training samples, they can be classified into supervised learning, semi-supervised learning and unsupervised learning algorithms.

In supervised learning, samples have two parts: one is the input features and the other is the output labels (Wang et al., 2016; Chen & Liu, 2018). Therefore, the input features come to be the causes and the output labels will be the effects. The purpose of supervised learning is to infer a functional relationship based on the training data that generalises well to the test data. The relationship is described as a set of equations and numerical coefficients.

In unsupervised learning, a set of features is available and observational information is missing for each sample (Chen & Liu, 2018). The features are caused by a set of hidden variables. The goal of unsupervised learning is to find relationships of the hidden variables in the observations. Some examples of unsupervised learning are based on clustering, blind source separation, density estimation, among others.

Semi-supervised learning lies between supervised and unsupervised learning (Kotsiantis et al., 2007). It uses labelled data (usually a few) and unlabelled data (usually many) during the training process. Semi-supervised learning algorithms were developed mainly because labelling data is very expensive or impossible in some applications. Examples of semi-supervised learning include semi-supervised classification and information recommender systems (Harrington, 2012).

Machine learning has many real-life applications. It is commonly used in banking (to detect fraudulent transactions (Fails et al., 2003), in finance (to predict stock prices (Shavlik et al., 1990), in marketing (to reveal consumer spending patterns (Janiesch, et al., 2021), and on the Internet (as part of search engines (Harrington, 2012). In biomedicine, MYCIN was proposed in the early 1970s at Stanford University.

This book will initially deal with data manipulation using dataframes in the Python programming language, embedded in the Jupyter-lab framework and Pandas. Data will then be extracted from csv files to be managed in dataframes. Indexing, selection and allocation, indexing in Pandas, tag-based selection, conditional selection, data allocation, summary functions, maps and grouping and sorting. Finally, we will proceed to program basic supervised learning models such as linear regression with a single variable, linear regression with multiple variables, saving and loading the training model, data management with dummy variables, separation of the training and test datasets.

Working with DataFrames

1.1. Preliminary installations

Download Python Python 3.8.2, from the following URL

https://www.python.org/downloads/release/python-382/

Install it.
Once Python is installed, perform the following instructions
Enter the CMD command line
Add the following command line

```
pip install jupyterlab
```

Next, install Jupyter Notebook,
Add the following command line

```
pip install notebook
```

Once installed, we proceed to install Pandas

To do so, follow the instructions below
Go to the following link
https://docs.conda.io/en/latest/miniconda.html

Download miniconda for Python 3.8, as shown in Figure 1.

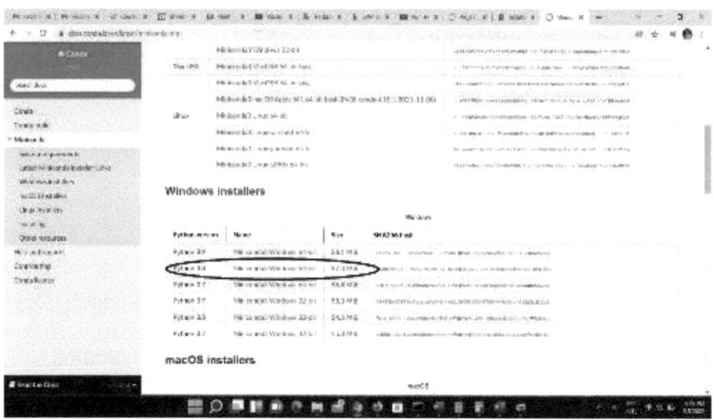

Figure 1. Mini-wave installation

Once you have installed all of the above in cmd type the following command line to launch Jupyter lab

```
jupyter-lab
```

1.2. Getting started with Dataframes

Open a new NoteBook, as shown in figure 2.

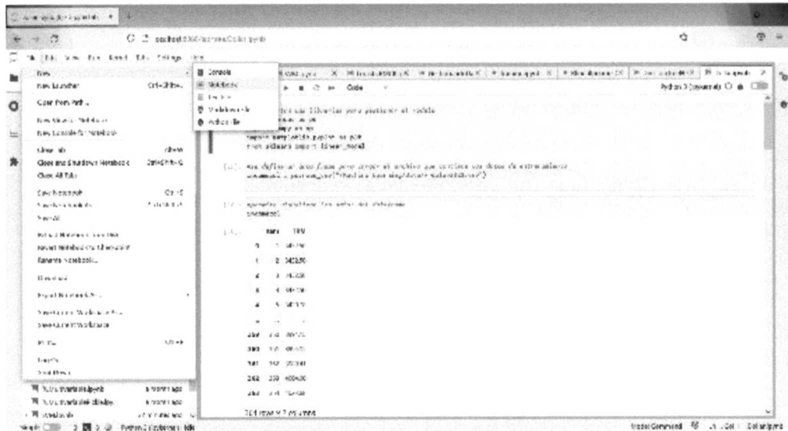

Figure 2. Creating a new Notebook project in Jupyterlab.

To work on ML in Python, it is necessary to give credits to (Pedregosa et al, 2011), for their valuable contributions to the different implemented Machine Learning algorithms, you can also access the Web Side[1] .

Continuing with the development, follow the instructions below:

1.3. Creating, reading and writing data in panda

Import pandas libraries

```
import pandas as pd
```

Creating data

1.3.1. DataFrame

A DataFrame is a table, which contains an array of individual data, each data has an individual value, each entry corresponds to a record of rows and columns.

For example

```
pd.DataFrame({'Yes': [100, 200], 'No': [50, 30]})
```

[1] https://scikit-learn.org/stable/about.html#citing-scikit-learn

Result

	Yes	No
0	100	50
1	200	30

DataFrame entries are not limited to integers, but can also include text strings, for example:

```
pd.DataFrame({'Jorge': ['Chicken.', 'Beef.', 'Pork.', 'Fish.'], 'Aida': ['Chicken.', 'Pork.', 'Vegetables.', 'Fruits.']})
```

Result

	Jorge	Aida
0	Chicken.	Chicken.
1	Beef.	Pork.
	Pork.	Vegetables.
	Fish.	Fruit.

We are using the constructor pd.DataFrame () to generate these objects

DataFrame. The syntax for declaring a new one is a dictionary whose keys are the column names Jorge and Aida in this example), and whose values are a list of entries. This is the standard way to construct a new DataFrame and the one you are most likely to encounter.

The dictionary list builder assigns values to the column labels, but only uses an ascending count from 0 (0, 1, 2, 3, ...) for the row labels. Sometimes this is fine, but often we want to assign these labels ourselves.

The list of row labels used in a DataFrame is known as an index. We can assign values to it using an index parameter in our constructor:

```
pd.DataFrame({'Jorge': ['Chicken.', 'Beef.', 'Pork.', 'Fish.'], 'Aida': ['Chicken.', 'Pork.', 'Vegetables.', 'Fruit.']},
         index=['Meal A', 'Meal B', 'Meal C', 'Meal D'])
```

Result

	Jorge	Aida
Food A	Chicken.	Chicken.
Food B	Beef.	Pork.
Food C	Pork.	Vegetables.
Food D	Fish.	Fruits.

1.3.2. Series

A series, on the other hand, is a sequence of data values. If a DataFrame is a table, a Series is a list. And, in fact, you can create one with nothing but a list:

```
pd.Series([1, 2, 3, 4, 5, 6, 7, 8, 9])
```

Result
```
0 1
1 2
2 3
3 4
4 5
5 6
6 7
7 8
8 9
dtype: int64
```

```
pd.Series([1.1, 2.2, 3.3, 4.4, 5.5, 6.6, 7.7, 8.8, 9.9])
```

```
0 1.1
1 2.2
2 3.3
3 4.4
4 5.5
5 6.6
6 7.7
7 8.8
8 9.9
dtype: float64
```

A Series is, in essence, a single column of a DataFrame. Therefore, you can assign column values to the Series in the same way as before, using an index parameter. However, a series does not have a column name, it just has a general name:

```
pd.Series([3.5, 4.0, 4.5, 4.0], index=['Quiz', 'Workshop', 'Partial', 'Final' ], name='Student Diana')
```

Result

Quiz 3.5
Workshop 4.0
Partial 4.5
Final 4.0
Name: Student Diana, dtype: float64

1.3.3. Read data files

Being able to create a DataFrame or a series by hand is useful. But, most of the time, we will not create our own data by hand. Instead, we will work with data that already exists.

Data can be stored in any of several different forms and formats. By far the most basic of these is the humble CSV file. When you open a CSV file, you get something like this:

So, a CSV file is a table of comma-separated values. Hence the name: "Comma Separated Values" or CSV.

Let's put our toy datasets aside now and see what a real dataset looks like when we read it into a DataFrame. We'll use the pd.read_csv() function to read the data into a DataFrame. It looks like this:

Sales January	Sales February	Sales March
130	201	
350	340	100
410	1110	2011
677	6000	5500
550		800
36900		5500

```
sales = pd.read_csv("/Machine Learning/Data/mydata_sales.csv".csv")
```

DataFrame using the head () command, which takes the first five rows:

```
sales.head()
```

Result

	Sales January	Sales February	Sales March
0	130	201	
1	350	340	100
	410	1110	2011
	677	6000	5500

the shape attribute checks how big the resulting DataFrame is:

```
sales.shape
```

Result
(6, 3)

Where it corresponds to the 6 rows and 3 columns described in the file

The pd.read_csv() function is well endowed, with more than 30 optional parameters that you can specify. For example, you can see in this dataset that the CSV file has a built-in index, which pandas did not automatically detect. To make pandas use that column for the index (instead of creating a new one from scratch), we can specify an index_col.

```
sales = pd.read_csv("/Machine Learning/Data/mydata_sales.csv", index_col=0)
sales.head()
```

	Sales February	Sales March
Sales January		
130	201	900
350	340	100
410	1110	2011
677	6000	5500
550		800

1.3.4. Indexing, selection and allocation

In Python, we can access the property of an object by accessing it as an attribute. A vehicle dealership object, for example, might have a property of city, make, month, week vehicle, quantity, unit price. Columns in a pandas DataFrame work in the same way.

Before starting, to run the exercise, create a Sales_vehicles.csv file with the following information

City	Brand	Month	Week	Vehicle	Quantity	Unit_price
Monteria	Toyota	January	1	Auto	105	80000
Monteria	Toyota	January	1	Truck	100	180000
Cerete	Suzuki	January	1	Omnibus	210	250000
Cerete	Toyota	February	1	Van	220	170000
Lorica	Suzuki	January	1	Motorbike	583	
Cerete	Renault	February		Auto		25000
Monteria	Renault	January		Truck	150	250000
Cienaga de Oro	Mazda	January		Omnibus	190	180000
Tierralta	Toyota	March		Van	255	180000
New town	Mazda	March		Motorbike	485	3000
Monteria	Suzuki	April		Auto		26000
New town	Chevrolet	March		Truck		256000
Pelayo	Ford	May		Omnibus		190000
Pelayo	Chevrolet	June		Van	223	170000
Monteria	Chevrolet	June		Motorbike		
Sahagun	Toyota	March		Auto	991	80000
Sahagun	Ford	January		Truck	990	221000
Lorica	Suzuki	May		Omnibus	25	177000
Lorica	Mazda	January		Van	80	77000
Tierralta	Mazda	February		Motorbike	544	

Execute the following instruction to load the file

car_sales = pd.read_csv("/Machine Learning/Data/Car_Sales.csv")

to access the city property can be used:

venta_vehiculos.head

Result
<bound method NDFrame.head of City Brand Month Month Week Vehicle Quantity
0 Monteria Toyota January 1 Auto 105
1 Monteria Toyota January 1 Truck 1
2 Cerete Suzuki January 1 Omnibus 2
3 Cerete Toyota February 1 Truck 2
4 Lorica Suzuki January 1 Moto 583
5 Cerete Renault February 2 Auto 121
6 Monteria Renault January 2 Truck 1
7 Cienaga de Oro Mazda January 2 Omnibus 1
8 Tierralta Toyota March 2 Truck 2
9 New Town Mazda March 2 Moto 485
10 Monteria Suzuki April 3 Auto 122
11 New Town Chevrolet March 3 Truck 1
12 Pelayo Ford Mayo 3 Omnibus 2
13 Pelayo Chevrolet June 3 Truck 2
14 Monteria Chevrolet June 3 Motorcycle 630
15 Sahagun Toyota March 4 Auto 99
16 Sahagun Ford January 4 Truck 990
17 Lorica Suzuki May 4 Omnibus 1
18 Lorica Mazda January 4 Pickup Truck 2
19 Tierralta Mazda Mazda February 4 Motorcycle 544>

If we have a Python dictionary, we can access its values using the indexing operator ([]). We can do the same with columns in a DataFrame:

venta_vehiculos.city

Result

0 Monteria
1 Monteria
2 Cerete
3 Cerete
4 Lorica
5 Cerete
6 Monteria
7 Cienaga de Oro
8 Tierralta
9 New town
10 Monteria
11 New town
12 Pelayo
13 Pelayo
14 Monteria
15 Sahagun
16 Sahagun
17 Lorica
18 Lorica
19 Tierralta
Name: City, dtype: object

If we have a Python dictionary, we can access its values using the indexing operator ([]). We can do the same with columns in a DataFrame:

venta_vehiculos['City']]

Result

0 Monteria
1 Monteria
2 Cerete
3 Cerete
4 Lorica
5 Cerete
6 Monteria
7 Cienaga de Oro
8 Tierralta
9 New town
10 Monteria

```
11 New town
12 Pelayo
13 Pelayo
14 Monteria
15 Sahagun
16 Sahagun
17 Lorica
18 Lorica
19 Tierralta
Name: City, dtype: object
```

These are the two ways to select a specific series from a DataFrame. Neither is more or less syntactically valid than the other, but the [] indexing operator has the advantage that it can handle column names with reserved characters in them (for example, if we had a country providence column, reviews.country providence wouldn't work).

vehicle_sale['City'][0]

Result

Monteria

1.3.5. Indexing in pandas

The indexing operator and attribute selection are nice because they work just like in the rest of the Python ecosystem. As a beginner, this makes them easy to learn and use. However, pandas has its own access operators, loc and iloc. For more advanced operations, these are the ones you are supposed to use.

Index-based selection

Panda indexing operates in one of two paradigms. The first is index-based selection: selecting data based on its numerical position in the data. iloc follows this paradigm.

To select the first row of data in a DataFrame, we can use the following:

sale_vehicles.iloc[0]

Result
Monteria City
Toyota Brand
Month January
Week 1
Vehicle Auto
Quantity 105
Name: 0, dtype: object

Both loc and iloc are the first row and the second column. This is the opposite of what we do in native Python, which is first column, second row.

This means that it is a bit easier to retrieve rows and a bit harder to retrieve columns.

To retrieve a column with iloc, we can do the following:

```
sale_vehicles.iloc[:,0]
```

Result

0 Monteria
1 Monteria
2 Cerete
3 Cerete
4 Lorica
5 Cerete
6 Monteria
7 Cienaga de Oro
8 Tierralta
9 New town
10 Monteria
11 New town
12 Pelayo
13 Pelayo
14 Monteria
15 Sahagun
16 Sahagun
17 Lorica
18 Lorica
19 Tierralta
Name: City, dtype: object

On its own, the: operator, which also comes from native Python, means "all". However, when combined with other selectors, it can be used to indicate a range of values. For example, to select the country column only from the first, second and third rows, we would do:

```
sale_vehicles.iloc[:4, 0]
```

Result

```
0 Monteria
1 Monteria
2 Cerete
3 Cerete
Name: City, dtype: object
```

Or, to select only the third and fifth entries, we would do:

```
sale_vehicles.iloc[2:5, 0]
```

Result

```
3 Cerete
4 Lorica
Name: City, dtype: object
```

It is also possible to pass on a list:

```
sale_vehicles.iloc[[0, 3, 5], 0]
```

Result

```
0 Monteria
3 Cerete
5 Cerete
Name: City, dtype: object
```

Finally, it is worth knowing that negative numbers can be used in the selection. This will start counting forward from the end of the values. So, for example, here are the last five elements of the dataset.

```
sale_vehicles.iloc[-6:]
```

Result

City	Brand	Month	Week	Vehicle	Quantity	Toyota	index_backwards
Monteria	Chevrolet	June		Motorbike		everyone	
Sahagun	Toyota	March		Auto	99	everyone	5
Sahagun	Ford	January		Truck	990	everyone	
Lorica	Suzuki	May		Omnibus	1	everyone	
Lorica	Mazda	January		Van		everyone	
Tierralta	Mazda	February		Motorbike	544	everyone	1

1.3.6. Tag-based selection

The second paradigm for attribute selection is the one followed by the loc operator: tag-based selection. In this paradigm, what matters is the value of the data index, not its position.

For example, to get the first entry in reviews, we would now do the following:

vehicle_sale.loc[0, 'City']]

Result
Monteria

iloc is conceptually simpler than loc because it ignores the indices of the dataset. When we use iloc, we treat the dataset as a large array (a list of lists), one in which we have to index by position. loc, on the other hand, uses the index information to do its job. Since your dataset generally has meaningful indexes, it is generally easier to do things using loc. For example, here is an operation that is much easier with loc:

sale_vehicles.loc[:, ['City', 'Make', 'Quantity']]]

Result

	City	Brand	Quantity
0	Monteria	Toyota	105

	City	Brand	Quantity
1	Monteria	Toyota	1
	Cerete	Suzuki	
	Cerete	Toyota	
	Lorica	Suzuki	583
5	Cerete	Renault	
	Monteria	Renault	1
	Cienaga de Oro	Mazda	1
8	Tierralta	Toyota	
	New town	Mazda	485
10	Monteria	Suzuki	
	New town	Chevrolet	1
	Pelayo	Ford	
	Pelayo	Chevrolet	
	Monteria	Chevrolet	
	Sahagun	Toyota	99
	Sahagun	Ford	990
	Lorica	Suzuki	1
	Lorica	Mazda	
	Tierralta	Mazda	544

1.3.7. Conditional selection

So far we have indexed several steps of data, using structural properties of the DataFrame itself. However, to do interesting things with the data, we often need to ask condition-based questions.

For example, suppose we are specifically interested in vehicles sold in Monteria.

We can start by checking whether each wine is Italian or not:

venta_vehiculos.ciudad == 'Monteria'.

Result
0 True
1 True
2 False
3 False
4 False

5 False
6 True
7 False
8 False
9 False
10 True
11 False
12 False
13 False
14 True
15 False
16 False
17 False
18 False
19 False
Name: City, dtype: bool

This operation produced a series of boolean true/false values based on the country of each record. This result can be used within loc to select the relevant data:

```
venta_vehiculos.loc[venta_vehiculos.Ciudad == 'Monteria']]
```

Result

	City	Brand	Month	Week	Vehicle	Quantity
0	Monteria	Toyota	January	1	Auto	105
1	Monteria	Toyota	January	1	Truck	1
	Monteria	Renault	January		Truck	1
10	Monteria	Suzuki	April		Auto	
	Monteria	Chevrolet	June		Motorbike	

We also wanted to know which ones are better than average, for example vehicles with sales above 70

We can use the ampersand (&) to join the two questions together:

```
venta_vehiculos.loc[(venta_vehiculos.Ciudad == 'Monteria') &
(venta_vehiculos.Cantidad >= 70)])
```

Result

	City	Brand	Month	Week	Vehicle	Quantity
0	Monteria	Toyota	January	1	Auto	105
10	Monteria	Suzuki	April		Auto	

	City	Brand	Month	Week	Vehicle	Quantity
	Monteria	Chevrolet	June		Motorbike	

Suppose we buy any wine made in Italy or with an above-average rating. For this we use a pipe (|):

venta_vehiculos.loc[(venta_vehiculos.Ciudad == 'Lorica') | (venta_vehiculos.Cantidad >= 50)])

Result

	City	Brand	Month	Week	Vehicle	Quantity
0	Monteria	Toyota	January	1	Auto	105
	Lorica	Suzuki	January	1	Motorbike	583
5	Cerete	Renault	February		Auto	
	New town	Mazda	March		Motorbike	485
10	Monteria	Suzuki	April		Auto	
	Monteria	Chevrolet	June		Motorbike	
	Sahagun	Toyota	March		Auto	99
	Sahagun	Ford	January		Truck	990
	Lorica	Suzuki	May		Omnibus	1
	Lorica	Mazda	January		Van	
	Tierralta	Mazda	February		Motorbike	544

Pandas comes with some built-in conditional selectors, two of which we will highlight here.

The first is isin. isin allows you to select data whose value "is in" a list of values. For example, this is how we can use it to select vehicles sold in Montería or Lorica:

venta_vehiculos.loc[venta_vehiculos.Ciudad.isin(['Monteria', 'Lorica'])])

Result

	City	Brand	Month	Week	Vehicle	Quantity
0	Monteria	Toyota	January	1	Auto	105
1	Monteria	Toyota	January	1	Truck	1

	City	Brand	Month	Week	Vehicle	Quantity
	Lorica	Suzuki	January	1	Motorbike	583
	Monteria	Renault	January		Truck	1
10	Monteria	Suzuki	April		Auto	
	Monteria	Chevrolet	June		Motorbike	
	Lorica	Suzuki	May		Omnibus	1
	Lorica	Mazda	January		Van	

The second is isnull (and its companion is not null). These methods allow you to highlight values that are (or are not) empty (NaN). For example, to filter out vehicles that lack a quantity label in the dataset, this is what we would do:

car_sale.loc[car_sale.quantity.notnull()].

Result

	City	Brand	Month	Week	Vehicle	Quantity
0	Monteria	Toyota	January	1	Auto	105
1	Monteria	Toyota	January	1	Truck	1
	Cerete	Suzuki	January	1	Omnibus	
	Cerete	Toyota	February	1	Van	
	Lorica	Suzuki	January	1	Motorbike	583
5	Cerete	Renault	February		Auto	
	Monteria	Renault	January		Truck	1
	Cienaga de Oro	Mazda	January		Omnibus	1
8	Tierralta	Toyota	March		Van	
	New town	Mazda	March		Motorbike	485
10	Monteria	Suzuki	April		Auto	
	New town	Chevrolet	March		Truck	1
	Pelayo	Ford	May		Omnibus	
	Pelayo	Chevrolet	June		Van	
	Monteria	Chevrolet	June		Motorbike	
	Sahagun	Toyota	March		Auto	99
	Sahagun	Ford	January		Truck	990
	Lorica	Suzuki	May		Omnibus	1
	Lorica	Mazda	January		Van	
	Tierralta	Mazda	February		Motorbike	544

1.3.8. Assign data

By doing the opposite, assigning data to a DataFrame is easy. You can assign a constant value:

```
vehicle_sale['Toyota'] = 'Toyota_Japan' = 'Toyota_Japan'.
sale_vehicles['Toyota']]
```

```
Result
0 Toyota_Japan
1 Toyota_Japan
2 Toyota_Japan
3 Toyota_Japan
4 Toyota_Japan
5 Toyota_Japan
6 Toyota_Japan
7 Toyota_Japan
8 Toyota_Japan
9 Toyota_Japan
10 Toyota_Japan
11 Toyota_Japan
12 Toyota_Japan
13 Toyota_Japan
14 Toyota_Japan
15 Toyota_Japan
16 Toyota_Japan
17 Toyota_Japan
18 Toyota_Japan
19 Toyota_Japan
Name: Toyota, dtype: object
```

Or with an iterable of values:

```
car_sale['index_backwards'] = range(len(car_sale), 0, -1)
venta_vehiculos['index_backwards']]
```

```
Result
0 20
1 19
2 18
3 17
4 16
5 15
6 14

8 12
9 11
10 10
11 9
12 8
13 7
14 6
```

```
15 5
16 4
17 3
18 2
19 1
Name: index_backwards, dtype: int64
```

1.3.9. Summary functions

Pandas provides many simple "summary functions" (not an official name) that restructure data in some useful way. For example, consider the describe() method:

```
venta_vehiculos.Cantidad.describe()
```

Result

```
count 20.000000
mean 184.800000
std 291.584528
min 1.000000
25% 1.750000
50% 2.000000
75% 212.750000
max 990.000000
Name: Quantity, dtype: float64
```

This method generates a high-level summary of the attributes of the given column. It is type-aware, which means that its output changes according to the type of the input data. The above output only makes sense for numeric data; for string data, this is what we get:

```
venta_vehiculos.Marca.describe()
```

Result

```
count 20
unique 6
top Toyota
freq 5
Name: Brand, dtype: object
```

If you want to get a particular simple summary statistic about a particular column in a DataFrame or Series, there is usually a useful pandas function that makes this possible.

For example, to see the average of the points assigned (e.g., what the average vehicle sales are), we can use the mean () function:

```
venta_vehiculos.Cantidad.mean()
```

Result
184.8

To see a list of unique values, we can use the unique() function:

```
venta_vehiculos.Marca.unique()
```

Result
array(['Toyota', 'Suzuki', 'Renault', 'Mazda', 'Chevrolet', 'Ford'],
    dtype=object)

To see a list of unique values and the frequency with which they occur in the dataset, we can use the value_counts () method:

```
venta_vehiculos.Marca.value_counts()
```

Result
Toyota 5
Suzuki 4
Mazda 4
Chevrolet 3
Renault 2
Ford 2
Name: Brand, dtype: int64

1.3.10. Maps

A map is a term, borrowed from mathematics, for a function that takes one set of values and "maps" them to another set of values. In data science, we often need to create new representations from existing data, or to transform data from the format it is in now to the format we want it to be in later. Maps are what handle this work.

There are two mapping methods that you will use frequently.

map () is the first and a bit simpler. For example, suppose we wanted to re-analyse vehicle sales. We can do this in the following way:

```
revision_vehicle_sale_mean = vehicle_sale.quantity.mean()
vehicle_sale.quantity.map(lambda p: p - vehicle_sale_revision_mean)

0 -79.8
1 -183.8
2 -182.8
3 -182.8
4 398.2
5 -63.8
6 -183.8
7 -183.8
8 -182.8
9 300.2
10 -62.8
11 -183.8
12 -182.8
13 -182.8
14 445.2
15 -85.8
16 805.2
17 -183.8
18 -182.8
19 359.2
Name: Quantity, dtype: float64
```

The function that passes map() must expect a single value from the Series (a point value, in the example above) and return a transformed version of that value. map() returns a new Series where all the values have been transformed by its function.

apply () is the equivalent method if we want to transform a complete DataFrame by calling a custom method on each row.

```
def remean_quantity(row):
```

```
    row.Quantity = row.Quantity - average_vehicle_sale_revision
    return row

venta_vehiculos.apply(remean_quantity, axis='columns')
```

	City	Brand	Month	Week	Vehicle	Quantity	Toyota	index_backwards
0	Monteria	Toyota	January	1	Auto	-79.8	Toyota_Japan	
1	Monteria	Toyota	January	1	Truck	-183.8	Toyota_Japan	
	Cerete	Suzuki	January	1	Omnibus	-182.8	Toyota_Japan	
	Cerete	Toyota	February	1	Van	-182.8	Toyota_Japan	
	Lorica	Suzuki	January	1	Motorbike	398.2	Toyota_Japan	
5	Cerete	Renault	February		Auto	-63.8	Toyota_Japan	
	Monteria	Renault	January		Truck	-183.8	Toyota_Japan	
	Cienaga de Oro	Mazda	January		Omnibus	-183.8	Toyota_Japan	
8	Tierralta	Toyota	March		Van	-182.8	Toyota_Japan	
	New town	Mazda	March		Motorbike	300.2	Toyota_Japan	
10	Monteria	Suzuki	April		Auto	-62.8	Toyota_Japan	10
	New town	Chevrolet	March		Truck	-183.8	Toyota_Japan	
	Pelayo	Ford	May		Omnibus	-182.8	Toyota_Japan	8
	Pelayo	Chevrolet	June		Van	-182.8	Toyota_Japan	
	Monteria	Chevrolet	June		Motorbike	445.2	Toyota_Japan	
	Sahagun	Toyota	March		Auto	-85.8	Toyota_Japan	5
	Sahagun	Ford	January		Truck	805.2	Toyota_Japan	
	Lorica	Suzuki	May		Omnibus	-183.8	Toyota_Japan	
	Lorica	Mazda	January		Van	-182.8	Toyota_Japan	
	Tierralta	Mazda	February		Motorbike	359.2	Toyota_Japan	1

If we had called reviews.apply () with axis = 'index', then instead of passing a function to transform each row, we would need to give a function to transform each column.

Note that map() and apply() return new and transformed Series and DataFrames, respectively. They do not modify the original data they are called on. If we look at the first row of reviews, we can see that it still has its original point value.

```
sale_vehicles.head(1)
```

	City	Brand	Month	Week	Vehicle	Quantity	Toyota	index_backwards
0	Monteria	Toyota	January	1	Auto	105	Toyota_Japan	

Pandas provides many common mapping operations built in. For example, here is a faster way to remean our column of points:

```
revision_vehicle_sale_mean = vehicle_sale.quantity.mean()
venta_vehiculos.Cantidad - revision_venta_vehiculos_media
```

```
Result
0 -79.8
1 -183.8
2 -182.8
3 -182.8
4 398.2
5 -63.8
6 -183.8
7 -183.8
8 -182.8
9 300.2
10 -62.8
11 -183.8
12 -182.8
13 -182.8
14 445.2
15 -85.8
16 805.2
17 -183.8
18 -182.8
19 359.2
Name: Quantity, dtype: float64
```

In this code we are performing an operation between many values on the left hand side (everything in the array) and a single value on the right hand side (the middle value). Pandas looks at this expression and realises that we should try to subtract that middle value from each value in the data set.

Pandas will also understand what to do if we perform these operations between series of equal length. For example, an easy way to combine city and vehicle make information in the dataset would be to do the following:

```
venta_vehiculos.Ciudad + " - " + venta_vehiculos.Marca
```

```
0 Monteria - Toyota
1 Monteria - Toyota
2 Cerete - Suzuki
3 Cerete - Toyota
4 Lorica - Suzuki
```

5 Cerete - Renault
6 Monteria - Renault
7 Cienaga de Oro - Mazda
8 Tierralta - Toyota
9 New Town - Mazda
10 Monteria - Suzuki
11 New town - Chevrolet
12 Pelayo - Ford
13 Pelayo - Chevrolet
14 Monteria - Chevrolet
15 Sahagun - Toyota
16 Sahagun - Ford
17 Lorica - Suzuki
18 Lorica - Mazda
19 Tierralta - Mazda
dtype: object

1.3.11. Grouping and sorting

Maps allow us to transform data into a DataFrame or Series one value at a time for an entire column. However, we often want to group our data and then do something specific to the group the data is in.

As you will learn, we do this with the groupby() operation. We'll also cover some additional topics, such as more complex ways to index your DataFrames, along with how to sort your data.

Group analysis

One function we have been using a lot so far is the value_counts() function. We can replicate what value_counts() does by doing the following:

```
venta_vehiculos.groupby('Cantidad').Cantidad.count()
```

Quantity
1 5
2 6
99 1
105 1
121 1

122 1
485 1
544 1
583 1
630 1
990 1
Name: Quantity, dtype: int64

groupby () created a group of reviews that assigned the same point values to the given wines. Then, for each of these groups, we took the points column () and counted how many times it appeared. value_counts () is just a shortcut to this groupby () operation.

We can use any of the summary functions we have used before with this data. For example, to get the cheapest vehicle in each point value category, we can do the following:

sale_vehicles.groupby('Quantity').Unit_price.min()

Quantity
2 190000
19 256000
25 177000
80 77000
100 180000
105 80000
121 25000
122 26000
150 250000
190 180000
210 250000
220 170000
223 170000
255 180000
485 3000
544 5000
583 1500
630 5000
990 221000
991 80000
Name: Unit_Price, dtype: int64

You can think of each group we generate as a portion of our DataFrame that contains only data with matching values. This DataFrame is accessible to us directly using the apply() method, and then we can manipulate the data in any way we see fit. For example, here is a way to select the name of the first vehicle reviewed for each brand in the dataset:

```
venta_vehiculos.groupby('Ciudad').apply(lambda df: df.Marca.iloc[0])
```

```
Result
City
Suzuki Cherry
Cienaga de Oro Mazda
Lorica Suzuki
Monteria Toyota
Pelayo Ford
New Mazda New Town
Sahagun Toyota
Tierralta Toyota
dtype: object
```

For even more detailed control, you can also group by more than one column. For example, this is how we would select vehicles and unit price:

```
venta_vehiculos.groupby(['City', 'Brand']).apply(lambda df:
df.loc[df.Unit_Price.idxmax()])
```

Result

		City	Brand	Month	Week	Vehicle	Quantity	Unit_price
City	**Brand**							
Cerete	**Renault**	Cerete	Renault	February		Auto		25000
	Suzuki	Cerete	Suzuki	January	1	Omnibus	210	250000
	Toyota	Cerete	Toyota	February	1	Van	220	170000
Cienaga de Oro	**Mazda**	Cienaga de Oro	Mazda	January		Omnibus	190	180000
Lorica	**Mazda**	Lorica	Mazda	January		Van	80	77000
	Suzuki	Lorica	Suzuki	May		Omnibus	25	177000
Monteria	**Chevrolet**	Monteria	Chevrolet	June		Motorbike		
	Renault	Monteria	Renault	January		Truck	150	250000
	Suzuki	Monteria	Suzuki	April		Auto		26000
	Toyota	Monteria	Toyota	January	1	Truck	100	180000
Pelayo	**Chevrolet**	Pelayo	Chevrolet	June		Van	223	170000
	Ford	Pelayo	Ford	May		Omnibus		190000

		City	Brand	Month	Week	Vehicle	Quantity	Unit_price
City	**Brand**							
New town	**Chevrolet**	New town	Chevrolet	March		Truck		256000
	Mazda	New town	Mazda	March		Motorbike	485	3000
Sahagun	**Ford**	Sahagun	Ford	January		Truck	990	221000
	Toyota	Sahagun	Toyota	March		Auto	991	80000
Tierralta	**Mazda**	Tierralta	Mazda	February		Motorbike	544	
	Toyota	Tierralta	Toyota	March		Van	255	180000

Another groupby() method worth mentioning is agg(), which allows you to run a bunch of different functions on your DataFrame simultaneously. For example, we can generate a simple summary statistic of the dataset as follows:

venta_vehiculos.groupby(['Marca']).Precio_Unitario.agg([len, min, max])

	len	min	max
Brand			
Chevrolet			256000
Ford		190000	221000
Mazda		3000	180000
Renault		25000	250000
Suzuki			250000
Toyota	5	80000	180000

1.3.12. Missing data types and values

Dtypes

The data type for a column in a DataFrame or a Series is known as dtype.

You can use the dtype property to take the type of a specific column. For example, we can get the dtype of the price column in the reviews DataFrame:

sale_vehicles.unit_price.dtype

Result
dtype('int64')

Alternatively, the dtypes property returns the dtype for each column in the DataFrame:

sale_vehicles.dtypes

Result
City object
Brand object
Month object
Week int64
Vehicle object
Quantity int64
Unit_price int64
dtype: object

The data types tell us something about how pandas store data internally. float64 means you are using a 64-bit floating point number; int64 means an integer of similar size, and so on.

One peculiarity to note (and shown very clearly here) is that columns consisting entirely of strings do not get their own type; instead, they are given the object type.

It is possible to convert a column of one type to another as long as the conversion makes sense using the astype() function. For example, we can transform the points column from its existing int64 data type to a float64 data type:

venta_vehiculos.Precio_Unitario.astype('float64')

Result
0 80000.0
1 180000.0
2 250000.0
3 170000.0
4 1500.0
5 25000.0
6 250000.0

Missing data

Missing values of entries are given the value NaN, short for "Not a number". For technical reasons, these NaN values are always of type d of float64.

Pandas provides some specific methods for missing data. To select NaN entries, you can use pd.isnull() (or its companion pd.notnull()). This is intended to be used like this:

car_sale[pd.isnull(car_sale.city)]]

City	Brand	Month	Week	Vehicle	Quantity	Unit_price
NaN	Toyota	February		Motorbike	600	8500

Replacing missing values is a common operation. Pandas provides a really useful method for this problem: fillna(). fillna() provides a few different strategies to mitigate such data. For example, we can simply replace each NaN with an "Unknown":

venta_vehiculos.Ciudad.fillna("Unknown")

```
Result
0 Monteria
1 Monteria
2 Cerete
3 Cerete
4 Lorica
5 Cerete
6 Monteria
7 Cienaga de Oro
8 Tierralta
9 New town
10 Monteria
11 New town
12 Pelayo
13 Pelayo
14 Monteria
15 Sahagun
16 Sahagun
17 Lorica
18 Lorica
19 Tierralta
20 Unknown
Name: City, dtype: object
```

1.3.13. Change of name and combination

Rename

The first function we will introduce here is rename(), which allows you to change index names and/or column names. For example, to change the points column in our dataset to score, we would do:

venta_vehiculos.rename(columns={'Quantity': 'Qty.Sold'})

Result

	City	Brand	Month	Week	Vehicle	Qty Sold	Unit_price
0	Monteria	Toyota	January	1	Auto	105	80000
1	Monteria	Toyota	January	1	Truck	100	180000
	Cerete	Suzuki	January	1	Omnibus	210	250000
	Cerete	Toyota	February	1	Van	220	170000
	Lorica	Suzuki	January	1	Motorbike	583	
5	Cerete	Renault	February		Auto		25000
	Monteria	Renault	January		Truck	150	250000
	Cienaga de Oro	Mazda	January		Omnibus	190	180000

rename () allows you to rename index or column values by specifying an index or column keyword parameter, respectively. It supports a variety of input formats, but generally a Python dictionary is the most convenient. Here is an example using it to rename some index items.

venta_vehiculos.rename(index={0: 'Primera_Entrada', 1: 'Segunda_Entrada'})

	City	Brand	Month	Week	Vehicle	Quantity	Unit_price
First_Entry	Monteria	Toyota	January	1	Auto	105	80000
Second_Entry	Monteria	Toyota	January	1	Truck	100	180000
	Cerete	Suzuki	January	1	Omnibus	210	250000
	Cerete	Toyota	February	1	Van	220	170000
	Lorica	Suzuki	January	1	Motorbike	583	
5	Cerete	Renault	February		Auto		25000

1.4. Linear regression model with a single variable

We wish to estimate the price of houses in the Los Ángeles neighbourhood in Montería, for which we have the following information (this data is fictitious), see table 1.

Table 1. House prices per square metre

area	price
98	275
115	370
125	410
150	470
175	520
200	580

Linear regression is one of the most famous techniques of describing data and making predictions about it (Uyanık, 2013; Chatterjee and Hadi, 2009; Kavitha et al.,2016). In image x the housing prices are shown, this information collected, will be used to predict the cost of real estate based on the values in table 1 in square meters.

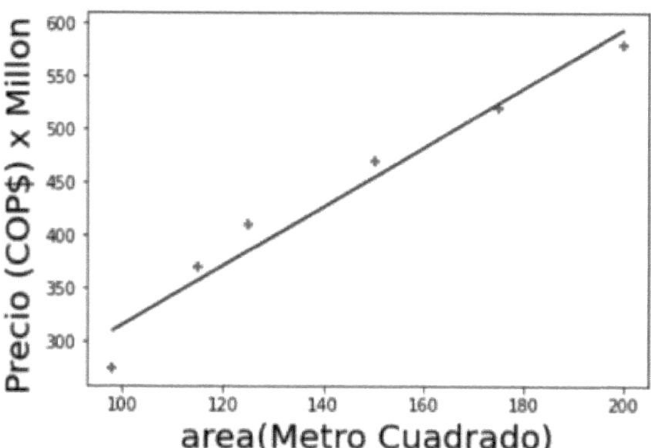

Figure 3. House prices per square metre

Given these reference prices in table 2, is it possible to calculate prices for houses with an area of 170 mts^2 and 250 mts^2 ?

Linear regression will help answer that question: it reduces your data into a line (the dotted one in the picture above), with the corresponding mathematical equation. If you know the equation of that line, you can find any output (y) given any input (x).

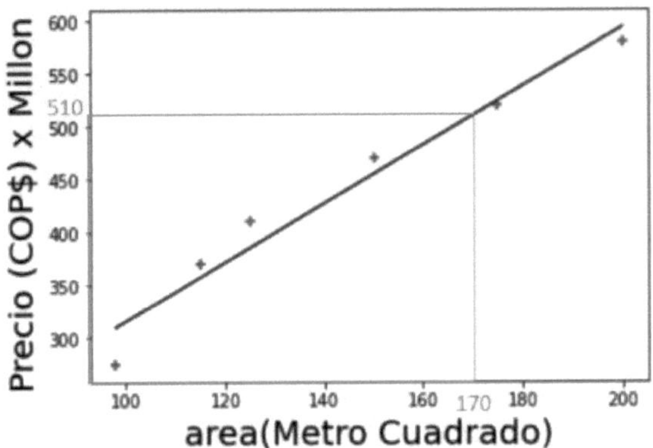

Figure 4. Predicted value for a 170 m house²

The initial data, in practice, is the so-called training dataset, i.e. house prices per square metre. The aim of the algorithm is to learn from this data to predict the prices of new houses. It is using input data to train the program. The training dataset can be summarised in a table 2, where the area corresponds to the length of the building (x) and the price corresponds to the value of the house (y).

Table 2. Training data (house prices per square metre)

area (x)	price (y)
98	275
115	370
125	410
150	470
175	520
200	580
****	****

The number of rows in table 2, the input variable x is the size of the individual dwelling in the left-hand column, and is the output variable, i.e. the price, in the right-hand column.

The table (x, y) shows a single example of generic training, i.e. if we want to specify in the training how much a house with an area of 115 mts^2 costs, the value of this would be 370 million Colombian pesos.

The equation that would be used to solve this problem would be[2]

$$y = mx + b$$

Where m is the slope or gradient

b is the intercept (y when x =0)

In other words

$$precio = m * area + b$$

The idea is to put all possible points in the Cartetian plane on a straight line, at a glance we can see that the blue line is the one that best fits the points, as shown in figure 5.

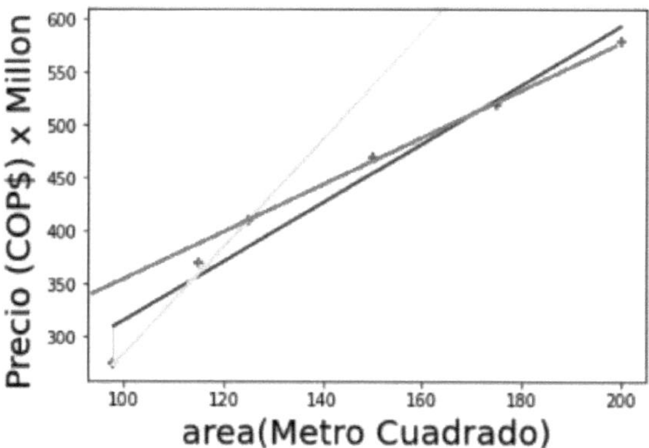

Figure 5. Line settings

[2] To learn more go to the following website https://mathsisfun.com/equation_of_line.html

To get the best fit, it is necessary to make use of the mean squared error (MSE) function, which measures how close a fitted line is to some data points. The lower the MSE, the closer the fit is to the data. Actually, there are many other functions that work well for such a task, but the MSE is the most commonly used for regression problems (Ohtani, 1996).

$$MSE = \frac{1}{n}\sum_{i=1}^{n}(y_i - \hat{y}_i)^2$$

To calculate the MSE, copy and paste the dataset from table 2 into the following link http://www.alcula.com/calculators/statistics/linear-regression/#gsc.tab=0

can validate the model before coding it

the result can be seen in figure 6.

Sample size: 6
Mean x (x̄): 143.83333333333
Mean y (ȳ): 437.5
Intercept (a): 38.141706744211
Slope (b): 2.776535063192
Regression line equation: y=38.141706744211+2.776535063192x

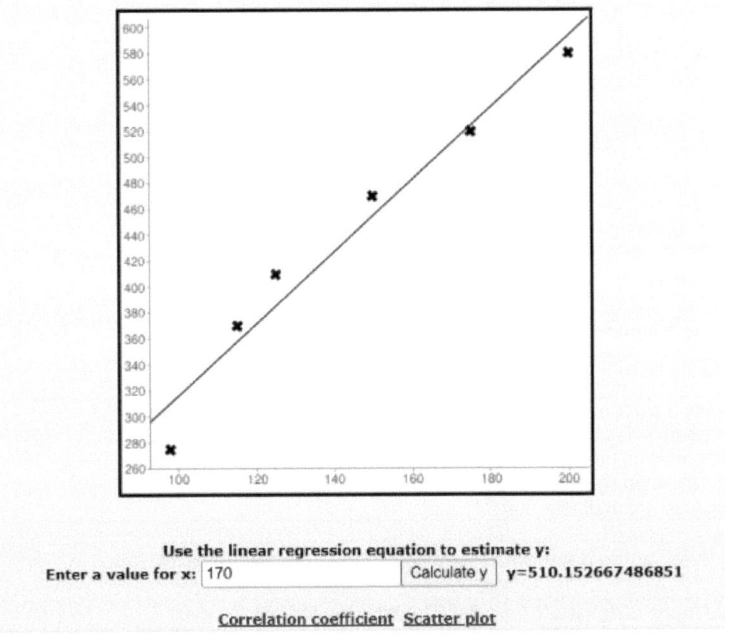

Figure 6. Validation of the model, from the website http://www.alcula.com/calculators/statistics/linear-regression/#gsc.tab=0.

With the information from the table xy create a CSV file with the name

precio_casas.csv

Then follow the instructions below as shown in figure 7.

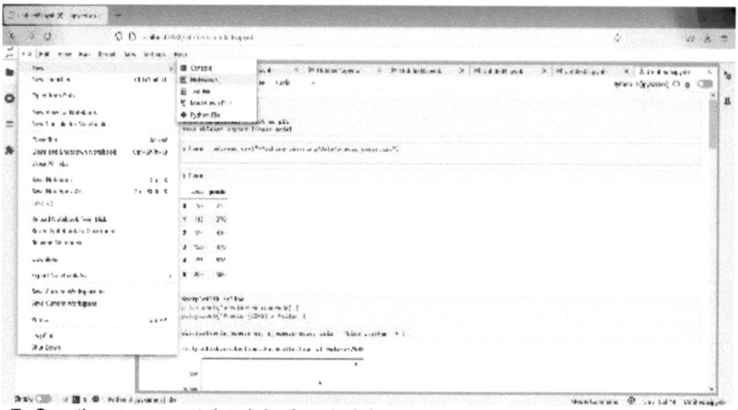

Figure 7. Creating a new notebook in Jupyterlab

Step-by-step

Enter the following instructions

```
# libraries are imported to manage the model
import pandas as pd
import numpy as np
import matplotlib.pyplot as plt
from sklearn import linear_model
```

Next, the s_home dataframe is defined to load the data from the file

```
#a data frame is defined to load the file containing the training data.
s_home = pd.read_csv("/Machine Learning/Data/price_home.csv")
```

The data is then displayed

```
#enables the dataframe data to be displayed
s_home
```

Result

	area	price
0	98	275
1	115	370
	125	410
	150	470
	175	520
5	200	580

The parameters for displaying the scatter plot are defined below.

```
%matplotlib inline
```

```
plt.xlabel('area(square metre)')
plt.ylabel('Price (COP$) x Million')

plt.scatter(s_home.area, s_home.price, colour ='blue',marker='+')
```

The result we obtain is Figure 8

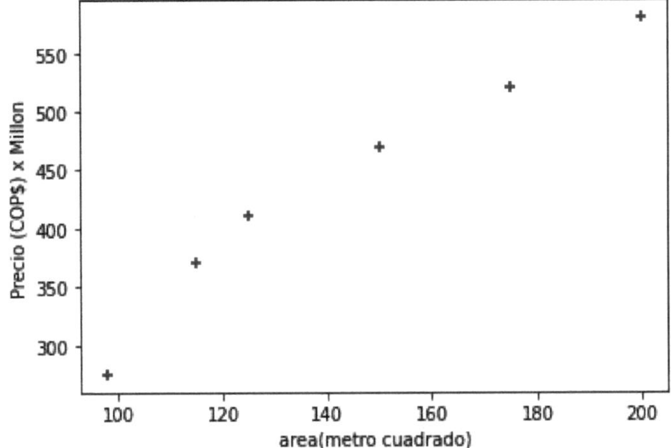

Figure 8. Scatter plot for houses based on square metre

Next, we pass the parameters to the linear regression model.

```
#The linear regression model is defined.
reg = linear_model.LinearRegression()
#The parameters of the model are set, which in this case are the area and the price.
reg.fit(s_home[['area']],s_home.price)
```

This results in

LinearRegression()

Next we will validate the model, for this we will put a value that is not in the training table.

```
#data to predict the value of the house according to the square meter
reg.predict([[170]])
```

As a result we have

array([510.15266749])

To test that everything is OK we will do a few tests

Recalling the formula

$$y = mx + b$$

```
#m is the slope or gradient
m=reg.coef_
m
```

The result is
array([2.77653506])

we will now calculate the intercept

```
#b is the intercept
b=reg.intercept_
b
```

The result of the intercept is

38.141706744211945

In the following, we will define the area

```
#area in square metres (x)
x=170
```

The result is

38.141706744211945

Now we get the value of y

```
#Finally, we obtain the value of y, which is the price of the house.
y=m*x+b
y
```

The result is

array([510.15266749])

As could be noticed, it is the same as the one obtained with the prediction function, and it is also the same as the one shown in figure 3, which was calculated to find the MSE.

We will then generate the complete graph with the straight line.

```
#The scatter plot and line are defined based on the prediction model, with area and price as parameters.
%matplotlib inline
plt.xlabel('area(Square Meter)',fontsize=20)
plt.ylabel('Price (COP$) x Million ',fontsize=20)
```

```
plt.scatter(s_home.area, s_home.price, colour ='red',marker='+')
plt.plot(s_home.area,reg.predict(s_home[[['area']])),colour='blue')
```

The result can be seen in figure 9.

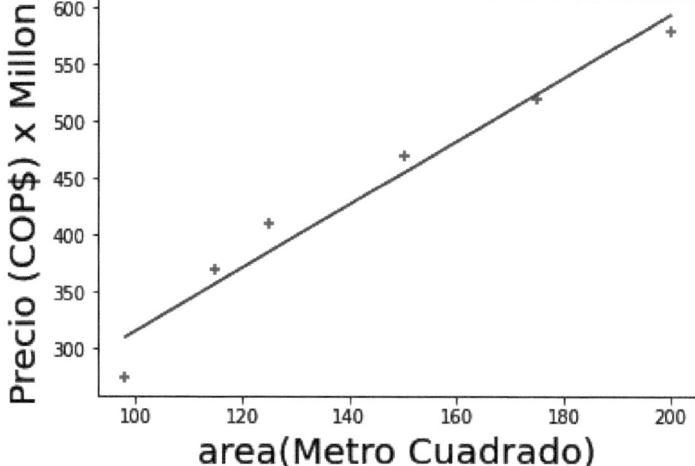

Figure 9. House prices per square metre

Once the model has been trained, we will proceed to test it with data from the area only.

To do this, build a csv file called area.csv with the following data

area
260

320
360

456
490
500
510
550
560
580
600

```
#A new dataframe is defined to load the area data.
sale_home2=pd.read_csv("/Machine Learning/Data/area.csv")
sale_home2.head(5)
```

The first 5 data will be displayed

	area
0	260
1	
	320
	360

Next, we will use the trained model to show us the prediction result with the area only.

```
reg.predict(sale_home2)
```

As a result we will have the value of the 14 houses according to their area.

```
array([ 760.04082317, 871.1022257 , 926.63292697, 1037.69432949,
       1148.75573202, 1304.24169556, 1398.64388771, 1426.40923834,
       1454.17458897, 1565.2359915 , 1593.00134213, 1648.5320434 ,
       1704.06274466, 2037.24695224])
```

If we want to display it in a more elegant way, we do the following

```
#We define a price variable to load the prediction data.
price=reg.predict(sale_home2)
```

```
#We assign the price based on the prediction to the sale_home dataframe.
sale_home2['price']=price
sale_home2
```

As a result we have

	area	price
0	2600	5.336644e+05
1	3000	5.879795e+05
	3200	6.151370e+05
	3600	6.694521e+05
		7.237671e+05
5	4560	7.998082e+05
	5490	9.260908e+05
	3490	6.545154e+05
8	3460	6.504418e+05
	4750	8.256079e+05
10	2300	4.929281e+05
	9000	1.402705e+06

	area	price
	8600	1.348390e+06
	7100	1.144709e+06

This information can be stored in a file called prediction.

```
sale_home2.to_csv("/Machine Learning/Data/Prediction.csv")
```

Figure 10 shows the file generated.

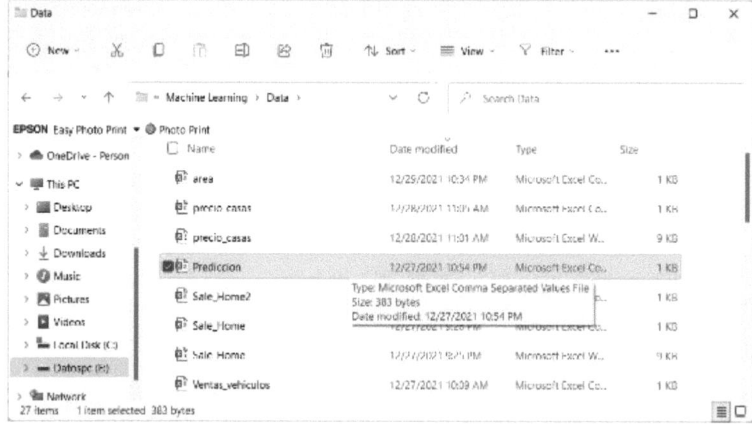

Figure 10. Prediction file generated.

This would be the complete code

```
# libraries are imported to manage the model
import pandas as pd
import numpy as np
import matplotlib.pyplot as plt
from sklearn import linear_model
#a data frame is defined to load the file containing the training data.
s_home = pd.read_csv("/Machine Learning/Data/price_home.csv")
#enables the dataframe data to be displayed
print(s_home)
#The scatter plot is defined
%matplotlib inline
plt.xlabel('area(square metre)')
plt.ylabel('Price (COP$) x Million')
plt.scatter(s_home.area, s_home.price, colour ='blue',marker='+')
#The linear regression model is defined.
reg = linear_model.LinearRegression()
reg.fit(s_home[[['area']],s_home.price)
#data to predict the value of the house according to the square meter
print('price ',reg.predict([[170]]), 'according to 170 sqm')
#m is the slope or gradient
```

```
m=reg.coef_
print('m: ',m)
#b is the intercept
b=reg.intercept_
print('b: ',b)
#area in square metres (x)
x=170
#Finally, we obtain the value of y, which is the price of the house.
y=m*x+b
y

print('price ',y)
#The scatter plot and line are defined based on the prediction model, with area and price as
parameters.
%matplotlib inline
plt.xlabel('area(Square Meter)',fontsize=20)
plt.ylabel('Price (COP$) x Million ',fontsize=20)
plt.scatter(s_home.area, s_home.price, colour ='red',marker='+')
plt.plot(s_home.area,reg.predict(s_home[[['area']])),colour='blue')
#A new dataframe is defined to load the area data.
sale_home2=pd.read_csv("/Machine Learning/Data/area.csv")
print(sale_home2.head(5))
print('prices',reg.predict(sale_home2))
#We define a price variable to load the prediction data.
price=reg.predict(sale_home2)
#We save the forecast file including square area and house prices.
sale_home2.to_csv("/Machine Learning/Data/Prediction.csv")
```

End of exercise.

1.5. Linear regression model with multiple variables

In multiple linear regression, there are p explanatory variables, and the relationship between the dependent variable and the explanatory variables is represented by the following equation:

$$Y = \beta_0 + \beta_1 x_{1i} + \beta_2 x_{2i} + .. + \beta_p x_{pi} + e_i$$

Where β_0

It is a constant term

β_1 to β_p are the coefficients relating the p explanatory variables to the variables of interest.

Multiple linear regression can be considered as an extension of simple linear regression, where there are p explanatory variables, or simple linear regression can be considered as a special case of multiple linear regression, where p = 1. The term "linear" is used because in multiple linear regression it is directly related to a linear combination of the explanatory variables.

In this opportunity it is desired to calculate the price of the flats in the neighbourhood Los Angeles in the city of Monteria, to take into account the area in square meters, the number of rooms and the age of construction of the flat, to determine the price of the same.

As can be seen below:

Table 3. Apartment prices

area	rooms	age	price
98		5	210
115		10	320
150	5	5	450
170			
200	5		800
220		5	
		10	
270	5	5	
280	5		1700
		5	1800
310		1	2100
350	8	5	2600
370	10	10	2700
380	8	5	2800
		5	3000
420	10	10	3100
440		5	3800

460	10		3900
		10	
500		10	4300
510		5	4500

With these reference prices, we would like to find a flat of 200 mts^2, 4 rooms and 10 years old. Similarly for a flat with 500 mts^2, 10 rooms and 15 years old.

According to the initial approach

$$Y = \beta_0 + \beta_1 x_{1i} + \beta_2 x_{2i} + .. + \beta_p x_{pi} + e_i$$

In other words, to find the price of the flat in question, the equation would have to be modified as follows

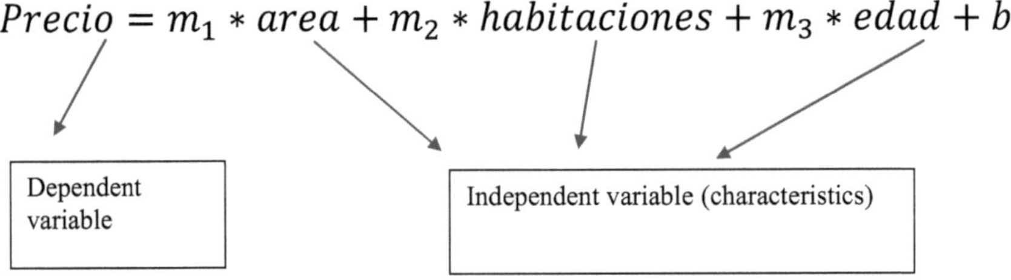

That is, in the form

$$y = m_1 x_1 + m_2 x_2 + m_3 x_3 + b$$

With the data provided in table 3, create a .csv file and name it apartment_the_angels.csv.

Next, in Jupiter-lab, open a new notebook

As shown in Figure 11

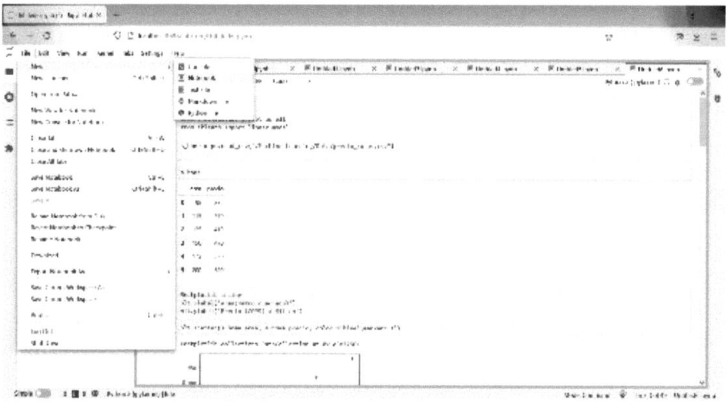

Figure 11. Creating a notebook for the linear regression problem with multiple variables

Then write the following lines of code

```
#A dataframe structure is created to load the data.
import pandas as pd
#This library allows the estimation of static models.
import statsmodels.api as sm
#This library adjusts the models that are used based on R-style formulas.
import statsmodels.formula.api as smf
# Allows to load the linear regression model
from sklearn import linear_model
```

click on the run button, as shown in figure 12.

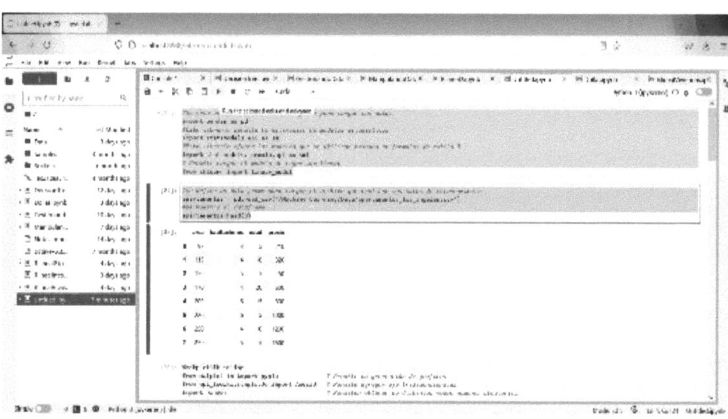

Figure 12. Executing command lines.

Then add the following lines

```
#a data frame is defined to load the file containing the training data.
flats = pd.read_csv("/Machine Learning/Data/apartamentos_los_angeles.csv")
```

```
#dataframe is displayed
flats.head(8)
```

click on the run button

the result will appear

	area	**rooms**	**age**	**price**
0	98		5	210
1	115		10	320
	150	5	5	450
	170			600
	200	5		800
5	220		5	
			10	
	270	5	5	

Next we are going to display a 3D graph, from the data loaded in the dataframe, add the following lines of code

```
%matplotlib inline
#This library allows the generation of graphics.
from matplotlib import pyplot
#This library allows to generate 3D graphics.
from mpl_toolkits.mplot3d import Axes3D


myfigure = pyplot.figure(figsize=(7, 5))
ax = Axes3D(myfigure)
#X-axis data
x1 = flats["area"]
#Y-axis data
x2 = flats["rooms"]
#Z-axis data (estimated price)
y = flats["price"]

ax.scatter(x1, x2, y, marker='*', c='r')
 # X-axis label
ax.set_xlabel('area in square metres')
# Y-axis label
ax.set_ylabel('# of rooms')
 # Z-axis label
ax.set_zlabel('price');
```

click on the run button and you will have a graph as shown in figure 13.

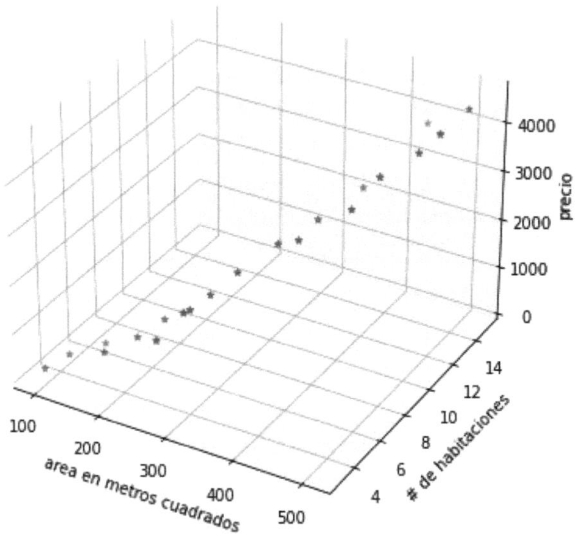

Figure 13. 3D graph of the data obtained in the dataframe

Next, we will adjust the model, add the following lines of code

```
# Model adjustment
fitted_model = smf.ols('price ~ area + rooms + age', data=apartments).fit()
print(fitted_model.summary())
```

click on the run button

it will look something like this

```
OLS Regression Results
==============================================================================
=
Variable dep.: price R-squared: 0.988
Model: OLS Adj. R-squared: 0.986
Method: Least Squares F-statistic: 461.6
Date: Mon, 03 Jan 2022 Prob (F-statistic): 1.76e-16
Time: 21:10:03 Log-Likelihood: -135.01
No. Observations: 21 AIC: 278.0
Df Residuals: 17 BIC: 282.2
Df Model: 3
Covariance Type: nonrobust
==============================================================================
===
coef std err t P>|t| [0.025 0.975].
------------------------------------------------------------------------------
Intercept -1197.2427 117.146 -10.220 0.000 -1444.398 -950.087
area 9.4061 0.687 13.699 0.000 7.957 10.855
rooms 53.0584 23.731 2.236 0.039 2.991 103.126
```

```
age 0.8185 7.496 0.109 0.914 0.914 -14.997 16.634
==============================================================================
=
Omnibus: 1.442 Durbin-Watson: 0.893
Prob(Omnibus): 0.486 Jarque-Bera (JB): 1.233
Skew: 0.537 Prob(JB): 0.540
Kurtosis: 2.492 Cond. No. 1.10e+03
==============================================================================
=

Notes:
[1] Standard Errors assume that the covariance matrix of the errors is correctly specified.
[2] The condition number is large, 1.1e+03. This might indicate that there are
strong multicollinearity or other numerical problems.
```

To corroborate the above information we write the following line of code

```
set_model.params
```

we click on run

it will look something like this

```
Intercept -1197.242717
area 9.406093
rooms 53.058376
age 0.818545
dtype: float64
```

As we can see, the highlighted values fit the fitted model.

Next, we will invoke the linear regression model fitted with the parameters of floor area, number of rooms and age of the flat.

```
#The linear regression model is defined.
reg=linear_model.LinearRegression()
#The fit of the linear regression model is requested based on the parameters of
#area, rooms and age, to calculate the price
reg.fit(flats[['area','rooms','age']],flats.price)
```

Click on the run button

It should look something like this

```
LinearRegression()
```

The coefficients are shown below

```
reg.coef_
```

Click run and it should look something like this

array([9.4060931 , 53.05837612, 0.81854508])

They are similar to those presented to the adjusted model in the upper part highlighted in lime green.

In the same way we validate the intercept

reg.intercept_

Click on run

It should look something like this

-1197.2427172738044

Equal to the value obtained from the fitted model.

Finally we will validate the model with the flat with a built area of 200 mts2, 4 rooms and 10 years of construction.

reg.predict([[200,4,10]])

The prediction value according to these parameters is

array([904.39485807])

now with 500 mts^2 , 10 rooms and 15 years old.

reg.predict([[500,10,15]])

The prediction value according to these parameters is
array([4048.66577033])

End of exercise

1.6. Saving and loading the training model

Once the training model has been created, it is not necessary to run the algorithm again to train it. For this purpose, there is a method called pickle, which allows the trained model to be stored and invoked when required. Figure 14 shows the operation and invocation of the model.

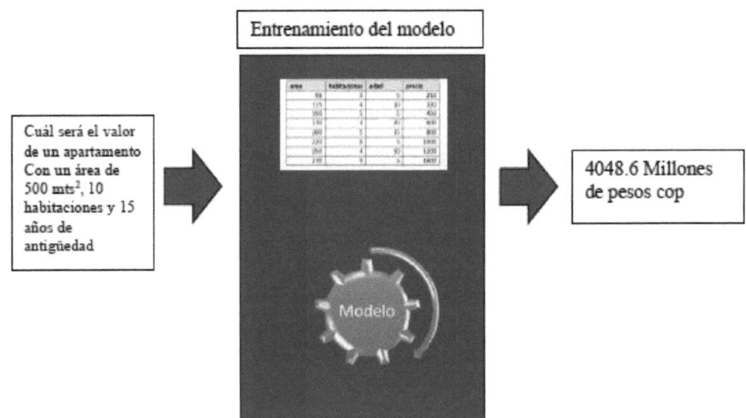

Figure 14. Training model for prediction

Then, in the same body of the multivariate linear regression notebook, add the following lines of code

```
#This library allows the serialisation and de-serialisation of object structures in Python.
import pickle
```

click on the run button

Then add the following lines of code

```
#This method allows to save the trained model file to the storage unit.
with open('model_pickle_LinearRegresion','wb') as f:
   pickle.dump(reg,f)
```

Click on the run button, to view the generated file go to Windows Explorer and check the model_pickle_LinealRegresion file, as shown in Figure 15.

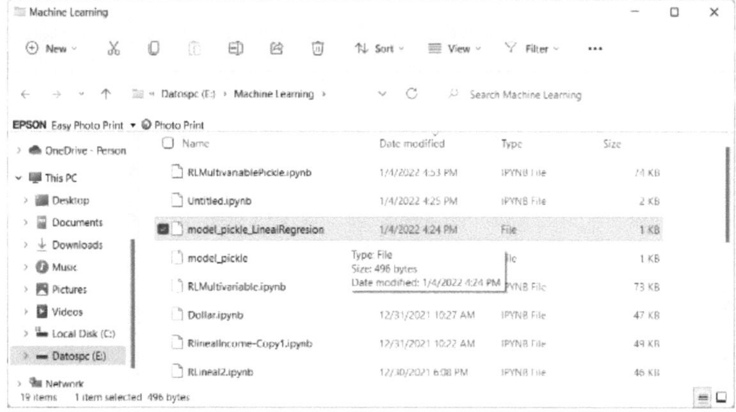

Figure 15. Generated training file

To use the trained model, open a new NoteBook.
Then add the following lines of code

```
#This library allows the serialisation and de-serialisation of object structures in Python.
import pickle
```

click on the run button

Then add the following lines of code

```
#This method allows you to open the almanac training model.
with open('model_pickle_LinearRegresion','rb') as f:
   mp= pickle.load(f)
```

Click on run

Then enter the parameters to the loaded model.

```
mp.predict([[500,10,15]])
```

Click on run

This will be the result

array([4048.66577033])

This result is similar to the one obtained when the model was tested.

End of exercise.

1.7. Data management with dummy variables or categorical variables (dummy)

Sometimes data are presented that are categorised as nominal and ordinal variables as shown in figure 16 below.

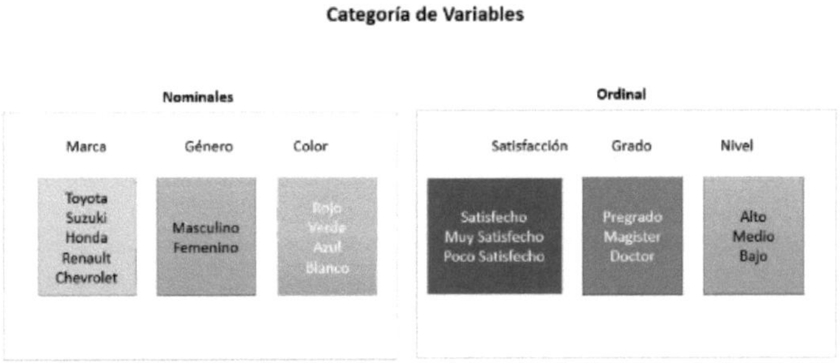

Figure 16. Categorisation of variables

The way pandas solves the problem is by assigning values to these nominal variables with dummy values, i.e. the variable name is replaced by a value of 0 or 1.

For example, in the table we see the original value of the data, where the nominal value is the mark and in table 4 the value of the mark is changed by the range of numbers from 1 to 4.

Table 4. Used vehicle prices

brand	model	price	age
Toyota Hilux	2006		
Toyota Hilux	2005	18.8	
Toyota Hilux	2004	17.7	
Toyota Fortuner	2018	90	
Toyota Fortuner	2019	100	
Toyota Fortuner	2016	70	5
Toyota Fortuner	2015		
Chevrolet Luv	2021		1
Chevrolet Luv	2020		
Chevrolet Luv	2019	73.3	
Chevrolet Luv	2018	64.9	
Ranault Logan	2019	30	
Ranault Logan	2018	28.2	
Ranault Logan	2017	26.5	5
Ranault Logan	2016	24.9	

Table 5. Used car prices by changing the brand name to values of (1 - 4)

brand	model	price	age
1	2006		
1	2005	18.8	
1	2004	17.7	
	2018	90	
	2019	100	
	2016	70	5
	2015		
	2021		1
	2020		
	2019	73.3	
	2018	64.9	
	2019	30	
	2018	28.2	
	2017	26.5	5
	2016	24.9	

Then, with the data supplied in table 5, create a csv file with the name of second_vehicles.csv

Then open a new NoteBook in Jupyter-Lab,

Add the following lines of code

```
#A dataframe structure is created to load the data.
import pandas as pd
#This library allows the estimation of statistical models.
import statsmodels.api as sm
#This library adjusts the models that are used based on R-style formulas.
import statsmodels.formula.api as smf
# Allows to load the linear regression model
from sklearn import linear_model
```

Click on the run button
Then add the following lines

```
#a data frame is defined to load the file containing the training data.
vehicles = pd.read_csv("/Machine Learning/Data/vehiculos_de_segunda.csv")
#dataframe is displayed
vehicles
```

Click on run

This will be the visualisation

	brand	model	price	age
0	Toyota Hilux	2006	20.0	
1	Toyota Hilux	2005	18.8	
	Toyota Hilux	2004	17.7	
	Toyota Fortuner	2018	90.0	
	Toyota Fortuner	2019	100.0	
5	Toyota Fortuner	2016	70.0	5
	Toyota Fortuner	2015	60.0	
	Chevrolet Luv	2021	83.0	1
8	Chevrolet Luv	2020	78.0	
	Chevrolet Luv	2019	73.3	
10	Chevrolet Luv	2018	64.9	
	Ranault Logan	2019	30.0	
	Ranault Logan	2018	28.2	
	Ranault Logan	2017	26.5	5
	Ranault Logan	2016	24.9	

Then, the variable brand will be changed to dummy

Add the following lines of code

```
#Converts categorical data into dummy variables or indicators
var_dummies=pd.get_dummies(vehicles.make)
var_dummies
```

Click on run

This will be the result

	Chevrolet Luv	Ranault Logan	Toyota Fortuner	Toyota Hilux
0	0	0	0	1
1	0	0	0	1
	0	0	0	1
	0	0	1	0
	0	0	1	0
5	0	0	1	0
	0	0	1	0
	1	0	0	0
8	1	0	0	0
	1	0	0	0
10	1	0	0	0
	0	1	0	0
	0	1	0	0
	0	1	0	0
	0	1	0	0

Each brand was assigned a column, the values of 1 in each column indicate the assigned values for each model and brand.

The values of the dataframe vehicles are then concatenated with the dummy variables.

Add the following lines of code

```
#concatenates the dataframe data with the dummy variables
fusion= pd.concat([vehicles,var_dummies],axis='columns')
merger
```

Click on the run button

It will look something like this

	brand	model	price	age	Chevrolet Luv	Ranault Logan	Toyota Fortuner	Toyota Hilux
0	Toyota Hilux	2006	20.0		0	0	0	1
1	Toyota Hilux	2005	18.8		0	0	0	1
	Toyota Hilux	2004	17.7		0	0	0	1
	Toyota Fortuner	2018	90.0		0	0	1	0
	Toyota Fortuner	2019	100.0		0	0	1	0
5	Toyota Fortuner	2016	70.0	5	0	0	1	0

	brand	model	price	age	Chevrolet Luv	Ranault Logan	Toyota Fortuner	Toyota Hilux
	Toyota Fortuner	2015	60.0		0	0	1	0
	Chevrolet Luv	2021	83.0	1	1	0	0	0
8	Chevrolet Luv	2020	78.0		1	0	0	0
	Chevrolet Luv	2019	73.3		1	0	0	0
10	Chevrolet Luv	2018	64.9		1	0	0	0
	Ranault Logan	2019	30.0		0	1	0	0
	Ranault Logan	2018	28.2		0	1	0	0
	Ranault Logan	2017	26.5	5	0	1	0	0
	Ranault Logan	2016	24.9		0	1	0	0

Then, the tick column will be removed to make the training and the respective prediction.

Add the following lines of code

```
#Eliminates the column mark
final=fusion.drop(['brand'], axis='columns')
definitive
```

Click on the run button

It will look something like this

	model	price	age	Chevrolet Luv	Ranault Logan	Toyota Fortuner	Toyota Hilux
0	2006	20.0		0	0	0	1
1	2005	18.8		0	0	0	1
	2004	17.7		0	0	0	1
	2018	90.0		0	0	1	0
	2019	100.0		0	0	1	0
5	2016	70.0	5	0	0	1	0
	2015	60.0		0	0	1	0
	2021	83.0	1	1	0	0	0
8	2020	78.0		1	0	0	0
	2019	73.3		1	0	0	0
10	2018	64.9		1	0	0	0
	2019	30.0		0	1	0	0
	2018	28.2		0	1	0	0
	2017	26.5	5	0	1	0	0
	2016	24.9		0	1	0	0

As can be seen, only the columns with numerical values were left for the respective training.

X will have the independent variables, so the price column must be removed.

Then add the following lines of code

```
#X is assigned the indpentity variables
X=definitive.drop('price',axis='columns')
X
```

Click on the run button

Only the value of the independent variables will be obtained, as shown below.

	model	age	Chevrolet Luv	Ranault Logan	Toyota Fortuner	Toyota Hilux
0	2006		0	0	0	1
1	2005		0	0	0	1
	2004		0	0	0	1
	2018		0	0	1	0
	2019		0	0	1	0
5	2016	5	0	0	1	0
	2015		0	0	1	0
	2021	1	1	0	0	0
8	2020		1	0	0	0
	2019		1	0	0	0
10	2018		1	0	0	0
	2019		0	1	0	0
	2018		0	1	0	0
	2017	5	0	1	0	0
	2016		0	1	0	0

Similarly, the value of the dependent variable price will be assigned in y

Add the following lines of code

```
#a and assigns the dependent variable, which in this case is price.
y=fusion.price
y
```

Click on the run button

Only prices will be shown

```
0 20.0
1 18.8
2 17.7
3 90.0
4 100.0
5 70.0
6 60.0
7 83.0
8 78.0
9 73.3
10 64.9
11 30.0
12 28.2
13 26.5
14 24.9
Name: price, dtype: float64
```

This will be followed by the training.

Add the following lines of code

```
#The linear regression model is defined.
reg=linear_model.LinearRegression()
reg.fit(X,y)
```

Click on the run button

It will look something like this

```
LinearRegression()
```

The prediction test will then be carried out.

Add the following line of code

```
#Can predict the value of the vehicle
reg.predict([[2005,16,0,0,0,1]])
```

Click on the run button

The result will be the predicted price for the 2005 model year Toyota Hilux, 16 years old.

```
array([18.84375])
```

Finally, to find out the confidence level of the prediction, add the following line of code

```
#Confidence level of the coefficient
reg.score(X,y)
```

The result is

```
0.975256502234059
```

End of exercise.

1.8. Separation of training and test datasets

Table 6 shows the following Dataset containing the data set for both training and testing. At first glance it is not clear which data will be used for training and which for testing. Obviously this is at the discretion of the data analyst.

Table 6. Dataset of used vehicles

kilometres	age	price
111044		73800
56327		139400
91732	5	107010
36210		164000
74030		129150
94951	5	109675
83686	5	131200
115872		79130
146450	8	49200
107826		90200

Once the data analyst decides which data to use for training and testing, as shown in Figure 17.

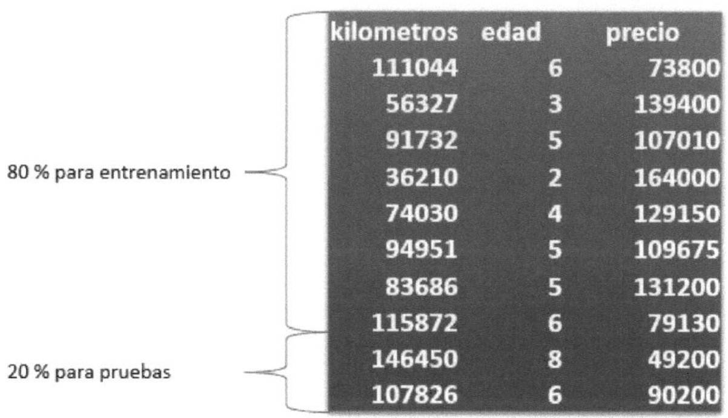

Figure 17. Dataset of used vehicles.

Pandas offers a simple way to separate training and training data.

With the data of table 6 create a csv file with the name carros_usados.csv

Then, open a new NoteBook in JupyterLab

Add the following lines of code

```
#A dataframe structure is created to load the data.
import pandas as pd
# Allows to load the linear regression model
from sklearn import linear_model
```

Click on the run button

Next, add the following lines of code

```
#a data frame is defined to load the file containing the training data.
vehicles = pd.read_csv("/Machine Learning/Data/cars_used.csv")
#dataframe is displayed
vehicles
```

Click on the run button

The dataset will be displayed

	kilometres	age	price
0	111044		73800
1	56327		139400
	91732	5	107010
	36210		164000
	74030		129150
5	94951	5	109675
	83686	5	131200
	115872		79130
8	146450	8	49200
	107826		90200

Below, a pair of scatter plots will be displayed to show mileage vs age and mileage vs price.

Add the following lines of code

```
#It is visualised in a scatter plot age of use vs. kilometres
import matplotlib.pyplot as plott
%matplotlib inline
plott.xlabel('age of use')
plott.ylabel('trip mileage')
plott.scatter(vehicles['age'],vehicles['kilometres'])
```

Click on the run button

Figure 18 will be displayed

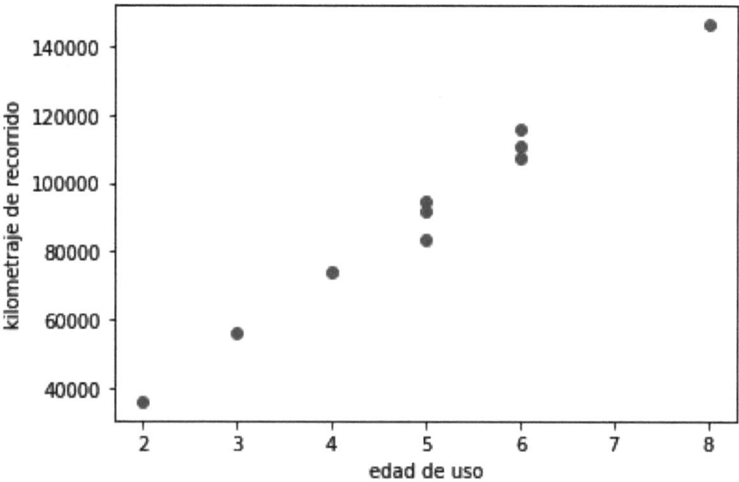

Figure 18. Scatterplot of used vehicles (Mileage vs. age of use)

Then add the following lines of code

```
#Price vs. kilometres is displayed in a scatter graph.
import matplotlib.pyplot as plott
%matplotlib inline
plott.xlabel('trip mileage')
plott.ylabel('price')
plott.scatter(vehicles['kilometres'],vehicles['price'])
```

Click on the run button

Figure 19 will be displayed.

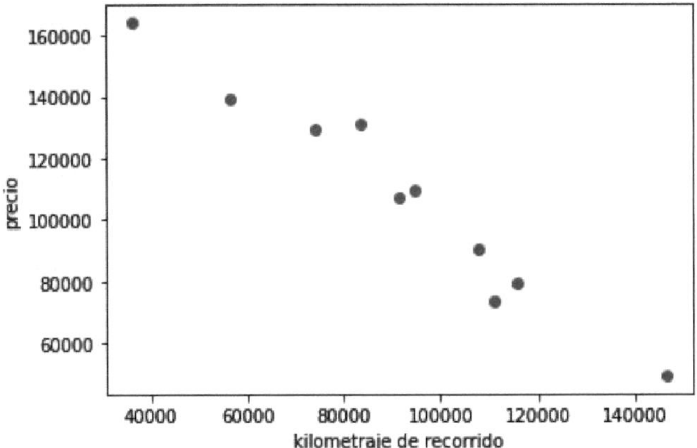

Figure 19. Scatterplot of used vehicles (Price Vs Mileage)

Next, add the following lines of code

```
#X separates the independent variables, i.e. vehicle mileage and vehicle age
X=vehicles[[['kilometres','age']]]
#In and assigns the dependent variable, which in this case is the price of the vehicle.
y=vehicles['price']]
#Independent variables are displayed
print(X)
```

Click on the run button

The data set of the independent variables will be displayed.

```
  kilometres age
0 111044 6
1 56327 3
2 91732 5
3 36210 2
4 74030 4
5 94951 5
6 83686 5
7 115872 6
8 146450 8
9 107826 6
```

Then add the following lines of code

```
#Dependent variable is displayed
print(y)
```

Click on the run button

The dependent variable is shown

```
0 73800
```

```
1 139400
2 107010
3 164000
4 129150
5 109675
6 131200
7 79130
8 49200
9 90200
Name: price, dtype: int64
```

Next, separate the training and test data, by adding the following lines of code

```
#train_test_split allows the separation of training and test data.
from sklearn.model_selection import train_test_split
#The data is divided into training and tests, furthermore the percentage for tests is defined by test_size=0.2.
#You can also randomly define which data will be used for training by random_state=10.
X_train, X_test, y_train, y_test= train_test_split(X,y,test_size=0.2, random_state=10)
```

Click on the run button

In the following, we will show values of the independent and dependent variables according to the separation

Add the following lines of code

```
#a variable tmX is defined to show how much data was taken from the dataset to train on the independent variables.
tmX=len(X_train)
print(tmX)
```

Click on the run button and the number of data used for training the independent variables will be displayed.

```
8
```

Next, we will do the same for the dependent variable, by adding the following lines of code

```
#a variable tmy is defined to show how much data was taken from the dataset to train on the dependent variable.
tmy=len(y_train)
print(tmy)
```

Click on the run button

The number of data used for and

```
8
```

Next, add the following lines of code

```
#a variable tmtestX is defined to show how much data was taken from the dataset to do the training of the independent variables.
tmtestX=len(X_test)
print(tmtestX)
```

Click on the run button

The number of data for the tests of the independent variables are shown.

Then add the following lines of code

```
#a variable tmtesty is defined to show how much data was taken from the dataset to test the dependent variable.
tmtesty=len(y_test)
print(tmtesty)
```

Click on the run button

The number of data used for the dependent variable is shown.

In the following, we will show the data selected for the training of the independent variables.
Add the following lines of code

```
#The training dataset stored in X_train is displayed.
print(X_train)
```

Click on the run button

The training data for the independent variables are shown.

```
  kilometres age
5 94951 5
6 83686 5
3 36210 2
1 56327 3
0 111044 6
7 115872 6
4 74030 4
9 107826 6
```

The training will then be followed by

Add the following lines of code

```
#The linear regression model is defined.
```

```
reg=linear_model.LinearRegression()
reg.fit(X_train,y_train)
```

Click on the run button

Next, add the following lines of code

```
#Allows prediction of vehicle values according to test data
reg.predict(X_test)
```

Click on the run button

The value of the predictions for the test data set will be displayed.

```
array([ 50535.2305584 , 107655.02352579])
```

Then add the following lines of code

```
#Data used for testing are shown
y_test
```

Click on the run button

The test data for the independent variable will be displayed.

```
8 49200
2 107010
Name: price, dtype: int64
```

Finally, we will show the level of accuracy of the prediction.

Add the following lines of code

```
#Allows to know the percentage of accuracy of the test data
reg.score(X_test,y_test)
```

Click on the run button

The result shows an accuracy level of

```
0.9986840822507688
```

End of exercise

REFERENCES

✓ Ohtani, K. (1996). On an adjustment of degrees of freedom in the minimim mean squared error ertimator. *Communications in Statistics--Theory and Methods*, *25*(12), 3049-3058.

✓ Kavitha, S., Varuna, S., & Ramya, R. (2016, November). A comparative analysis on linear regression and support vector regression. In *2016 online international conference on green engineering and technologies (IC-GET)* (pp. 1-5). IEEE.

✓ Uyanık, G. K., & Güler, N. (2013). A study on multiple linear regression analysis. *Procedia-Social and Behavioral Sciences*, *106*, 234-240.

✓ Chatterjee, S., & Hadi, A. S. (2009). *Sensitivity analysis in linear regression* (Vol. 327). John Wiley & Sons.

✓ Suzuki, S., N. Tsurusaki, and Y. Kodama. 2006. Distribution of an endangered burrowing spider *Lycosa ishikariana* in the San'in Coast of Honshu, Japan (Araneae: Lycosidae). Acta Arachnologica 55: 79-86.

✓ Daniel Jurafsky & James H. Martin, 2021. Speech and Language Processing. Draft of December 29, 2021.

✓ Wood Tomas, 2021. Sigmoid Function. https://deepai.org/profile/woodthom

✓ Breiman, L., & Shang, N. (1996). Born again trees. *University of California, Berkeley, Berkeley, CA, Technical Report*, *1*(2), 4.

✓ Schölkopf, B., Luo, Z., & Vovk, V. (Eds.) (2013). Empirical inference: Festschrift in honor of Vladimir N. Vapnik. Springer Science & Business Media.

✓ Cortes, C., & Vapnik, V. (1995). Support vector machine. *Machine learning*, *20*(3), 273-297.

✓ Breiman, L. (2001). Random forests. *Machine learning*, *45*(1), 5-32.

✓ Carleo, G., Cirac, I., Cranmer, K., Daudet, L., Schuld, M., Tishby, N., ... & Zdeborová, L. (2019). Machine learning and the physical sciences. Reviews of Modern Physics, 91(4), 045002.

✓ Alpaydin, E. (2021). Machine learning. MIT Press.

✓ El Naqa, I., & Murphy, M. J. (2015). What is machine learning? In machine learning in radiation oncology (pp. 3-11). Springer, Cham.

✓ Wang, H., Lei, Z., Zhang, X., Zhou, B., & Peng, J. (2016). Machine learning basics. Deep learning, 98-164.

✓ Chen, Z., & Liu, B. (2018). Lifelong machine learning. Synthesis Lectures on Artificial Intelligence and Machine Learning, 12(3), 1-207.

✓ Fails, J. A., & Olsen Jr, D. R. (2003, January). Interactive machine learning. In Proceedings of the 8th international conference on Intelligent user interfaces (pp. 39-45).

✓ Shavlik, J. W., Dietterich, T., & Dietterich, T. G. (Eds.) (1990). Readings in machine learning. Morgan Kaufmann.

✓ Janiesch, C., Zschech, P., & Heinrich, K. (2021). Machine learning and deep learning. Electronic Markets, 31(3), 685-695.

✓ Harrington, P. (2012). Machine learning in action. Simon and Schuster.

✓ Kotsiantis, S. B., Zaharakis, I., & Pintelas, P. (2007). Supervised machine learning: A review of classification techniques. Emerging artificial intelligence applications in computer engineering, 160(1), 3-24.

✓ Pedregosa, F., Varoquaux, G., Gramfort, A., Michel, V., Thirion, B., Grisel, O., ... & Duchesnay, E. (2011). Scikit-learn: Machine learning in Python. the Journal of machine Learning research, 12, 2825-2830.

Printed by Books on Demand GmbH, Norderstedt / Germany